GLOBAL NUCLEAR ENERGY PARTNERSHIP

GLOBAL NUCLEAR ENERGY PARTNERSHIP

ALAN N. BERNSTEIN
EDITOR

Nova Science Publishers, Inc.
New York

Copyright © 2009 by Nova Science Publishers, Inc.

All rights reserved. No part of this book may be reproduced, stored in a retrieval system or transmitted in any form or by any means: electronic, electrostatic, magnetic, tape, mechanical photocopying, recording or otherwise without the written permission of the Publisher.

For permission to use material from this book please contact us:
Telephone 631-231-7269; Fax 631-231-8175
Web Site: http://www.novapublishers.com

NOTICE TO THE READER

The Publisher has taken reasonable care in the preparation of this book, but makes no expressed or implied warranty of any kind and assumes no responsibility for any errors or omissions. No liability is assumed for incidental or consequential damages in connection with or arising out of information contained in this book. The Publisher shall not be liable for any special, consequential, or exemplary damages resulting, in whole or in part, from the readers' use of, or reliance upon, this material. Any parts of this book based on government reports are so indicated and copyright is claimed for those parts to the extent applicable to compilations of such works.

Independent verification should be sought for any data, advice or recommendations contained in this book. In addition, no responsibility is assumed by the publisher for any injury and/or damage to persons or property arising from any methods, products, instructions, ideas or otherwise contained in this publication.

This publication is designed to provide accurate and authoritative information with regard to the subject matter covered herein. It is sold with the clear understanding that the Publisher is not engaged in rendering legal or any other professional services. If legal or any other expert assistance is required, the services of a competent person should be sought. FROM A DECLARATION OF PARTICIPANTS JOINTLY ADOPTED BY A COMMITTEE OF THE AMERICAN BAR ASSOCIATION AND A COMMITTEE OF PUBLISHERS.

LIBRARY OF CONGRESS CATALOGING-IN-PUBLICATION DATA
Available upon request.

ISBN 978-1-60456-973-5

Published by Nova Science Publishers, Inc. ✣ *New York*

CONTENTS

Preface	vii
Results in Brief	1
Background	5
DOE's Original Engineering-Scale Approach Would Meet GNEP's Objectives If Advanced Recycling Technologies Are Successfully Developed	11
DOE's Accelerated Approach Would Likely Rely on Technologies That Fall Short of Meeting GNEP's Objectives	21
Conclusions	31
Recommendations for Executive Action	33
Agency Comments and Our Evaluation	35
Appendix I. Scope and Methodology	39
Appendix II. DOE's Use of Technology Readiness Levels to Assess the Maturity of Spent Fuel Recycling Technologies	43
Index	47

PREFACE

The Department of Energy (DOE) proposes under the Global Nuclear Energy Partnership (GNEP) to build facilities to begin recycling the nation's commercial spent nuclear fuel. GNEP's objectives include reducing radioactive waste disposed of in a geologic repository and mitigating the nuclear proliferation risks of existing recycling technologies. DOE originally planned a small engineering-scale demonstration of advanced recycling technologies being developed by DOE national laboratories. While DOE has not ruled out this approach, the current GNEP strategic plan favors working with industry to demonstrate the latest commercially available technology in full-scale facilities and to do so in a way that will attract industry investment. DOE has funded four industry groups to prepare proposals for full-scale facilities. DOE officials expect the Secretary of Energy to decide on an approach to GNEP by the end of 2008. GAO evaluated the extent to which DOE would address GNEP's objectives under (1) its original engineering scale approach and (2) the accelerated approach to building full-scale facilities. GAO analyzed DOE plans and industry proposals and interviewed DOE and industry officials concerning the pros and cons of both approaches.

RESULTS IN BRIEF

DOE's original approach to the domestic component of GNEP—building engineering-scale facilities—would meet GNEP's objectives if the advanced spent nuclear fuel recycling technologies on which it focused can be successfully developed and commercialized. Successful development of the advanced technologies would have a greater long-term impact, compared with existing technologies, on GNEP's waste reduction objective because the advanced technologies would recycle not only plutonium but also other radioactive elements that a geologic repository has limited capacity to accommodate. Keeping plutonium mixed with these other elements would mitigate proliferation risks relative to existing technologies because the mixture would be more difficult to steal or divert and to fashion into a nuclear weapon than pure plutonium. However, DOE's engineering-scale approach had two shortcomings. First, it lacked industry participation, potentially reducing the prospects for eventual commercialization of advanced recycling technologies. In particular, DOE's original approach included managing some of the radioactive waste separated from spent fuel in a way that would add to the cost and difficulty of operating a reprocessing plant while creating waste management challenges; recent industry proposals under DOE's accelerated approach include potentially less costly and complex alternatives for managing this waste. Second, while building an advanced reactor and R&D facility would allow DOE to conduct R&D that existing facilities have limited capability to support, DOE's schedule called for building an engineering-scale reprocessing plant before developing recycled fuel and other recycling technologies that would help determine design specifications for the plant. The schedule unnecessarily increased the risk that the plant would separate the materials in spent fuel in a form not suitable for recycling.

DOE's accelerated approach of building commercial-scale facilities would likely require using unproven evolutions of existing technologies that would reduce radioactive waste and mitigate proliferation risks to a much lesser degree than anticipated from more advanced technologies. In addition, this approach would likely require significant government investment. Two of the four industry consortia that received funding under GNEP proposed evolutions of existing technologies that recycle plutonium and uranium as mixed oxide, or MOX, fuel in existing reactors even though the GNEP strategic plan ruled out MOX technologies. Such technologies would reduce the quantity of high-level radioactive waste requiring geologic disposal to a much lesser degree than the advanced technologies envisioned under DOE's original approach. The evolutionary MOX technologies would also mitigate proliferation risks to a lesser degree because a plutonium-uranium mixture for recycling into MOX fuel would not contain other radioactive elements that would be recycled using more advanced technologies—elements that could pose barriers to obtaining pure plutonium for weapons. While the evolutionary technologies could allow DOE to accelerate recycling of spent nuclear fuel, fully meeting GNEP's waste reduction and nonproliferation objectives would require a later transition to more advanced technologies. The other two industry consortia proposed to address GNEP's waste and nonproliferation objectives by using technologies that are no more mature or in some cases less mature than the advanced technologies DOE had deemed appropriate for engineering-scale demonstration under its original approach. Thus, these proposals do not meet a goal of DOE's accelerated approach of working with industry: to avoid the need for engineering-scale facilities and increase the speed of arriving at commercial facilities. Under any of the industry proposals, DOE is unlikely to meet its goal of developing commercial-scale facilities in a way that will not require a large amount of government investment. For example, our review of all four industry proposals and interviews with DOE officials indicate that none of the consortia have proposed a way to pay for the initial advanced reactor other than through government funding. DOE officials acknowledge the limitations of the department's accelerated approach but cite other benefits, such as the potential to exert international influence on nonproliferation issues. They have also said that, if DOE pursues evolutionary MOX technologies, the department will only do so as part of a plan for a later transition to more advanced technologies.

Because DOE can fully address GNEP's waste reduction and nonproliferation objectives only by developing advanced technologies that are not yet ready for commercial deployment, we recommend that DOE reassess its preference for an accelerated approach to implementing GNEP. If DOE decides to pursue design and construction of engineering-scale facilities, we further recommend that DOE work with industry in doing so and defer building an engineering-scale reprocessing plant until conducting sufficient testing and development of recycled fuel to ensure that the output of the reprocessing plant is suitable for recycling.

We presented a draft of this report to DOE and NRC for comment. DOE agreed with many of our findings and concurred with our recommendations, directed toward the department's original engineering-scale approach to GNEP, to revise its schedule for an engineering-scale reprocessing plant and to work with industry to the extent possible. With regard to our recommendation that DOE reassess its preference for an accelerated approach to implementing GNEP, DOE stated that the department will continue to perform analyses to support the Secretary of Energy's decision on the direction for GNEP. DOE and NRC also provided detailed technical comments, which we have incorporated into our report as appropriate.

BACKGROUND

GNEP is part of the administration's Advanced Energy Initiative for reducing the nation's reliance on foreign sources of energy and increasing energy supplies in ways that protect the environment. The initiative seeks, among other things, to increase funding for R&D to enable the generation of more electricity from nuclear energy. Benefits of nuclear energy cited by the administration include avoidance of air pollution and greenhouse gas emissions, sufficient North American uranium reserves to fuel nuclear power plants for the foreseeable future and thus contribute to energy security, and the relatively low cost to operate nuclear power plants once they have been built and paid for. Under GNEP, the administration seeks to address two of nuclear energy's drawbacks—the need to dispose of spent nuclear fuel and the risk of nuclear proliferation.

DOE's Office of Nuclear Energy has primary responsibility for GNEP and has established asteering group to coordinate its implementation. The steering group includes other DOE offices with responsibility for programs related to GNEP, such as the Office of Civilian Radioactive Waste Management and the National Nuclear Security Administration, a separately organized agency within DOE that has responsibility for the department's nuclear nonproliferation programs. DOE national laboratories contributing to development of GNEP technologies include Argonne, Brookhaven, Idaho, Lawrence Berkeley, Lawrence Livermore, Los Alamos, Oak Ridge, Pacific Northwest, Sandia, and Savannah River. The Office of Nuclear Energy directed the Idaho National Laboratory to establish a technical integration office to serve as a point of contact with the other laboratories; integrate R&D and technology development activities; collect, analyze, and integrate financial and schedule data; and perform other administrative functions. The technical integration office oversees seven technical campaigns responsible for specific aspects of GNEP, each headed at a national laboratory by a campaign manager.

As required by the department's project management guidance, the Office of Nuclear Energy is currently evaluating alternative approaches and recycling scenarios for implementing GNEP's domestic component. Recycling scenarios differ by the technologies used and materials in spent fuel that are recycled. Such differences impact the degree to which GNEP's objectives would be addressed— for example, by the degree to which recycling of spent nuclear fuel would extend the technical capacity of a geologic repository to accommodate the remaining high-level radioactive waste (or, conversely, by the number of geologic repositories needed to dispose of the waste). DOE has estimated that, without recycling spent fuel, as many as four repositories could be required by 2100, assuming that nuclear energy maintains its current level of electricity generation and each additional repository has a limit of 70,000 metric tons. Even more repositories would be needed if, as DOE hopes, nuclear energy increases its share of the nation's electricity generation beyond the current level of 20 percent. In contrast, DOE hopes to develop advanced recycling technologies that would result in needing only one geologic repository this century.

Absent a second repository, DOE would not legally be able to avail itself of the Yucca Mountain geologic repository's full technical capacity unless the Nuclear Waste Policy Act of 1982 were amended. The act allows no more than 70,000 metric tons of spent fuel, or the high-level radioactive waste that results from reprocessing no more than 70,000 metric tons of spent fuel, to be disposed of in the repository unless a second repository is in operation. In contrast, GNEP is based on the assumption that the repository has a technical capacity to accommodate the high-level radioactive waste from reprocessing a much greater amount of spent fuel—if DOE is successful in developing advanced recycling technologies.[1] According to an analysis conducted by Argonne National Laboratory, the repository's technical capacity would be based on performance specifications designed to limit releases of the radioactivity in spent nuclear fuel to the environment. Since the spent nuclear fuel and other high-level waste stored in the repository can generate heat for long periods of time and the repository's performance can be affected by temperature, many of the performance specifications would be in the form of temperature limits. DOE proposes under GNEP to recycle or otherwise manage the materials in spent fuel that are

---

[1] References in this report to the capacity of the Yucca Mountain geologic repository are to its technical capacity unless otherwise noted.

significant contributors to decay heat, thereby allowing more of the remaining waste to be disposed of in the repository without exceeding the temperature limits.

Materials in Spent Nuclear Fuel

Hundreds of fuel assemblies—bundles of long metal tubes filled with uranium pellets—form the core of a typical nuclear power reactor (see fig. 1). Reactors produce energy when uranium atoms split (fission) into smaller elements, called fission products. Some of the uranium atoms do not split but rather are transformed into transuranics—elements heavier than uranium—such as plutonium. With the buildup of fission products, the uranium loses its ability to sustain a nuclear reaction, and the fuel assemblies are then removed for replacement. Removed assemblies (spent nuclear fuel) are some of the most hazardous materials made by humans. Without protective shielding, radiation from the spent fuel can kill a person directly exposed to it within minutes or increase the risk of cancer in people exposed to smaller doses.

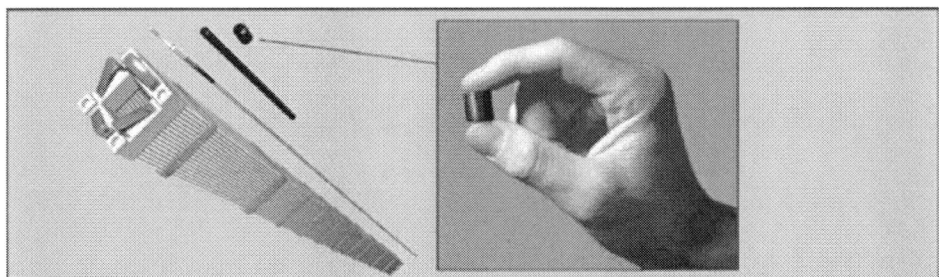

Source: Nuclear Energy Institute.

Figure 1. Nuclear Fuel Assembly and Uranium Pellet.

The uranium, fission products, and transuranics in spent fuel differ in terms of the impact they have on the technical capacity of a geologic repository as a result of their decay heat. They also differ in terms of their energy value and potential to be recycled. Uranium that was present in fresh fuel forms up to about 96 percent of the material in spent fuel. The uranium is not highly radioactive and contributes little to the decay heat of spent fuel. DOE has proposed under GNEP that the uranium, which has not lost all of its energy value, be stored for future recycling if it becomes economically viable to do so. Alternatively, DOE has suggested the

Table 1. Materials in Spent Nuclear Fuel and Their Potential Disposition under GNEP

Material	Percentage of spent fuel[a]	Decay heat characteristics of significance to the technical capacity of a geologic repository	Potential disposition under GNEP
Uranium	96	Uranium is not a significant contributor to decay heat in spent fuel.	Storage for later recycling or, if not recycled, disposal as low-level waste.
Fission products (e.g., cesium and strontium)	3	Cesium and strontium dominate decay heat for decades after spent fuel is removed from a reactor.	Disposal as high-level waste in a geologic repository, or, in the case of cesium and strontium, potential storage to allow radioactive decay to low-level waste.
Transuranics (plutonium, neptunium, americium, and curium)	1	Transuranics dominate decay heat for thousands of years after spent fuel is removed from a reactor.	Recycling in an advanced reactor, assuming successful R&D on recycled fuel containing the transuranics.
Total	100		

Source: GAO analysis of DOE information.
[a]The percentages of materials in spent fuel vary depending on the characteristics of particular fuel assemblies and do not include the structural hardware of the assemblies.

uranium could be managed as low-level radioactive waste, which does not require disposal in a geologic repository. Fission products, which constitute about 3 percent to 5 percent of the material in spent fuel, do not have energy value as fuel for a reactor and under GNEP would be disposed of as high-level radioactive waste. Two key fission products—cesium and strontium—are significant contributors to the decay heat in spent fuel. Because the fission products would no longer be contained within a fuel assembly, other ways of containing them would need to be used to ensure their safe disposal. DOE has conducted and continues to conduct R&D to enable disposal of the fission products as high-level waste. Transuranics, which include plutonium, constitute the smallest percentage of spent fuel. They are of primary interest under GNEP because they have energy value if advanced technologies for recycling them in reactors can be successfully developed. Transuranics also contribute to the long-term decay heat in spent fuel, and recycling them could extend the capacity of a geologic repository to accommodate the remaining high-level radioactive waste. (See table 1.)

Technologies for Recycling Spent Nuclear Fuel

Recycling spent fuel requires that a reprocessing plant break apart the used fuel assemblies and separate the reusable materials from the remaining waste. The reusable materials are then fabricated into recycled fuel for reactors. Under GNEP, DOE national laboratories are conducting R&D to develop advanced technologies for each of these steps. PUREX— the reprocessing technology originally developed in the United States to obtain plutonium for nuclear weapons and now used for commercial purposes in France, Japan, and other countries— separates out plutonium. According to DOE, the PUREX reprocessing technology can be adapted to recombine plutonium with uranium before the plutonium leaves the plant's radioactive processing area and thereby reduce the possibility of using a reprocessing plant to produce plutonium. Japan has made such an adaptation to its reprocessing plant. In contrast, advanced reprocessing technologies being developed by the national laboratories (generally known as the UREX+ suite of processes) would completely avoid separating out plutonium and would instead keep it mixed with one or more of the other transuranics, which would provide a higher level of proliferation resistance. The inclusion of other transuranics is intended to make it easier to detect theft or diversion of plutonium and to increase the difficulty of using the plutonium in a nuclear weapon.

The DOE national laboratories are also developing advanced technologies for fabricating and using recycled fuel that contains not only uranium and plutonium but also one or more of the other transuranics. In contrast, recycled fuel derived from existing technologies—called mixed oxide (MOX) fuel[2]—contains uranium and plutonium but not other transuranics, which are disposed of as waste despite their potential energy value. Under GNEP, DOE is considering various options for the recycled fuel and has not made a final decision on many of them, such as whether the reusable materials should be in the form of metal or oxide and whether all of the reusable materials should be fabricated together or some fabricated and recycled separately. Decisions on such options will in part affect

---

[2] Specifically, MOX fuel contains a mixture of plutonium oxide and uranium oxide.

another set of decisions on advanced reprocessing technologies (e.g., the technology chosen from among the UREX+ suite of processes).

The advanced reactor envisioned under GNEP would be used to transmute transuranics, or convert them into materials that generate decay heat for a shorter period of time, thereby extending the capacity of a geologic repository to store the remaining waste. The type of advanced reactor DOE plans to develop under GNEP is a "fast" reactor, as opposed to a "thermal" reactor. These terms refer to the neutron energy level at which a nuclear reaction is sustained in a reactor: Fast reactors operate with higher energy neutrons than thermal reactors. DOE specifically selected a fast reactor cooled by sodium as the advanced reactor for GNEP, in part because the technology for sodium-cooled fast reactors is considered to be more advanced than the technology for other types of fast reactors. However, while the United States and other countries have built and operated sodium-cooled fast reactors, largely for research purposes, no fast reactors are currently operating in the United States. In contrast, almost all commercial nuclear power plants and other operating reactors are thermal reactors—particularly light water reactors, which use ordinary water as a coolant.

NRC would have licensing and regulatory authority to ensure the safety of any commercial facilities for recycling spent fuel, including reprocessing plants and advanced reactors. Based on a preliminary assessment, NRC has concluded that changes in regulations and associated regulatory guidance would be necessary to support an efficient and effective licensing review of commercial GNEP facilities. Reprocessing and recycling spent nuclear fuel would also produce low-level radioactive waste, potentially in large quantity, and gaseous waste products. According to NRC, disposal of such wastes would face multiple technical, legislative, and regulatory challenges that, while not insurmountable, would nonetheless be significant.

DOE's Original Engineering-Scale Approach Would Meet GNEP's Objectives If Advanced Recycling Technologies Are Successfully Developed

Successful development and commercialization of advanced recycling technologies envisioned under the engineering-scale approach would have a greater long-term impact, compared with existing technologies, on GNEP's waste reduction objective. The advanced technologies would also mitigate proliferation risks relative to existing technologies. However, the engineering-scale approach lacked industry participation, potentially reducing the prospects for eventual commercialization of advanced technologies. Furthermore, the approach included building an engineering-scale reprocessing plant before conducting R&D that could help determine the plant's design requirements. In contrast, building an advanced reactor and R&D facility would allow DOE to conduct R&D that existing DOE facilities have limited capability to support.

Successful Development of Advanced Recycling Technologies Would Be an Initial Step toward Greatly Extending the Capacity of a Geologic Repository

DOE's original approach to GNEP would demonstrate at an engineering scale advanced technologies for recycling all of the transuranics in spent nuclear fuel. Transuranics are the dominant contributors over the long term to the spent fuel's decay heat, which is a primary limiting factor in the amount of spent fuel that can be accommodated in a geologic repository. Thus, successful development and implementation of technologies for recycling the transuranics could greatly extend the capacity of a geologic repository to contain the remaining high-level radioactive waste. For example, according to a recent analysis conducted by DOE's Argonne National Laboratory, recycling the transuranics could result, under certain conditions, in an almost sixfold increase in the amount of remaining waste that could be accommodated in a geologic repository with a capacity limited by temperature considerations. While the precise impact of recycling the transuranics would depend on many factors, such as the recycling technologies' effectiveness, the potential waste benefit of not disposing of transuranics in a geologic repository is well recognized, and development of advanced technologies for transmuting them has been a focus of DOE's Advanced Fuel Cycle Initiative.

DOE has analyzed various advanced technologies, such as different types of reactors, for transmuting transuranics in spent nuclear fuel. While an engineering-scale demonstration of any one set of advanced technologies may require that DOE narrow its focus to the exclusion of potentially worthy alternatives, there is substantial technical support for choosing torecycle transuranics using a fast reactor, as DOE has proposed under GNEP. The choice of reactor is critical from the standpoint of addressing GNEP's waste reduction objective because reactors differ in their ability to recycle the transuranics. DOE specifically selected a fast reactor as the advanced reactor envisioned under GNEP because its properties theoretically enable it to recycle transuranics more efficiently than thermal reactors. For example, analyses conducted by DOE national laboratories indicate that, whereas thermal reactors would be able to recycle the transuranics at most about two times, fast reactors would be capable of recycling the transuranics repeatedly. Achieving the waste reduction benefit of not disposing of transuranics in a geologic repository would require multiple recycling passes because recycled fuel, like conventional fuel used in light water reactors, loses its ability to sustain a nuclear reaction and is thus spent before the transuranics in it are fully consumed. Other organizations that have cited the benefit of successfully developing fast reactors to recycle transuranics include the Nuclear Energy Agency, DOE's Nuclear Energy Research Advisory Committee, and the Electric Power Research Institute.[1] For example, a Nuclear Energy Agency report issued in 2006 stated that studies have repeatedly demonstrated that fast reactors are more efficient than light water reactors for recycling and transmuting transuranics.

The focus on developing fast reactors under DOE's original approach to GNEP is also justified whether they are used alone or in combination with other reactor types. Because of the ability of fast reactors to transmute transuranics, many scenarios for recycling transuranics include the use of a fast reactor as an essential component. For example, DOE's Oak Ridge National Laboratory has studied the possibility of transmuting some of the transuranics in light water reactors and other transuranics in fast reactors. Such scenarios may provide advantages, such as the ability to use existing reactors without needing to deploy as many fast reactors, initial models of which are expected to be more expensive than light water reactors to build and operate. The advantages of scenarios for recycling transuranics in a combination of reactor types would have to be weighed against the disadvantages, such as the increased requirement for R&D on two sets of recycling technologies.

Successful development of fast reactors, even given their ability to transmute transuranics, would only be an initial step toward achieving GNEP's waste reduction objective. Like any technologies developed for recycling spent nuclear fuel, fast reactors would require widespread use and many years of operation before significantly reducing the inventory of transuranics that would otherwise require disposal in a geologic repository. For example, according to a hypothetical scenario analyzed by Idaho National Laboratory, fast reactors would transmute only about one-quarter of the transuranics produced by nuclear power plants by the end of the century. The scenario assumes that nuclear energy and recycling of spent fuel would grow at a brisk pace: By the end of the century, nuclear power would increase its share of the nation's electricity generation from about 20 percent to about 33 percent, and fast reactors would account for about 17 percent of the electricity generated by nuclear power plants. The scenario also assumes that three reprocessing plants, each with a capacity of 2,000 metric tons per year, would need to start up between 2020 and 2080.

---

[1] The Nuclear Energy Agency is part of the Organisation for Economic Co-operation and Development, an intergovernmental organization of industrialized countries. The mission of the Nuclear Energy Agency includes providing assessments of nuclear energy policy. The Nuclear Energy Advisory Committee, formerly the Nuclear Energy Research Advisory Committee, provides advice to the DOE Office of Nuclear Energy on science and technical issues related to DOE's nuclear energy program. The Electric Power Research Institute conducts R&D on behalf of the electricity industry, including R&D on nuclear energy technologies.

Advanced Recycling Technologies Envisioned under DOE's Original Approach to GNEP Pose Lower Proliferation Risks Than Existing Recycling Technologies

While advanced technologies for recycling spent nuclear fuel would pose a greater risk of proliferation in comparison with direct disposal in a geologic repository, they would reduce the risk of proliferation relative to existing reprocessing technologies that separate out plutonium. Direct disposal of spent nuclear fuel in a geologic repository provides a higher level of protection against theft or diversion of plutonium and its subsequent use in a nuclearweapon than recycling because spent fuel assemblies are highly radioactive for many years, and plutonium cannot be obtained from them other than by reprocessing the spent fuel. In contrast, existing spent fuel recycling technologies increase the risk of proliferation by separating out plutonium, which could conceivably be stolen or diverted more easily than a large radioactive fuel assembly. Existing recycling facilities address this risk through high levels of security and safeguards technologies to detect theft or diversion of nuclear materials.

DOE's advanced recycling technologies offer the possibility of reducing— but not eliminating—the risk of proliferation relative to existing recycling technologies. The advanced reprocessing technologies that DOE is developing (the UREX+ suite of processes) would keep plutonium mixed with one or more of the other transuranics. Of these technologies, the one that DOE had identified as the preferred option under its original approach to GNEP (the UREX+1a process) would keep plutonium mixed with all of the other transuranics, the radiation of which could create a barrier to handling the plutonium mixture and fabricating it into a nuclear weapon. However, even with this radiation barrier, the risk of theft or diversion from a reprocessing plant would necessitate high levels of security and the use of safeguards technologies. For example, the Savannah River Site's engineering analysis of a commercial-scale reprocessing plant using DOE's advanced reprocessing technology found that nuclear materials in the plant would fall into a category requiring a high level of protection under DOE security standards. The risk of theft or diversion from an advanced reprocessing plant could be even higher if DOE designed the plant to use one of the other UREX+ processes, which generally keep plutonium mixed with fewer radioactive transuranics.

DOE's original approach would further address the risk of proliferation by developing advanced safeguards technologies, such as equipment capable of near real-time monitoring of materials being reprocessed, and testing them in the initial facilities proposed under GNEP, particularly the R&D facility. According to the

GNEP safeguards campaign manager, existing safeguards technologies are not capable on their own of meeting the standard for detecting plutonium diversion that DOE hopes to meet with advanced technologies. Furthermore, the use of advanced reprocessing technologies that keep plutonium mixed with other transuranics would require the development of new safeguards technologies capable of detecting and identifying not only plutonium but also other transuranics.

Lack of Industry Participation Could Reduce the Prospects for Commercialization and Widespread Use of Advanced Recycling Technologies

DOE's original approach to GNEP did not reflect the input of industry on how to commercialize advanced technologies for recycling spent nuclear fuel. In particular, DOE's original approach to GNEP included the proposal to manage two of the key fission products—cesium and strontium—in a way that some in industry have questioned as too ambitious. DOE had planned to develop advanced reprocessing technologies to separate cesium and strontium and to dispose of them separately from other high-level radioactive waste placed in a geologic repository. According to Argonne National Laboratory's analysis, separately disposing of cesium and strontium would multiply the capacity-extending effect of recycling transuranics. The analysis suggests that keeping cesium and strontium as well as the transuranics out of a geologic repository with a capacity limited by temperature considerations could result in about a 100-fold increase in the amount of remaining waste that could be accommodated. According to the analysis, separation and disposal of cesium and strontium would not, on its own, allow any increase in the amount of remaining waste that could be accommodated in a temperature-limited repository since the transuranics are the dominant contributors to decay heat over the long term.

Separation of cesium and strontium would nonetheless create waste management challenges while also increasing the cost and complexity of a reprocessing plant. Although the impact on the capacity of a repository could be dramatic, cesium and strontium would still need to be managed as radioactive waste while undergoing radioactive decay—for approximately 300 years according to DOE's estimate. DOE has suggested that a site for cesium and strontium could be located at the reprocessing plant. DOE national laboratory officials have suggested that an alternative to storing cesium and strontium at a reprocessing plant is to create a dedicated site at the Yucca Mountain repository.

However, the need to create such a site would entail challenges, such as public opposition. Furthermore, an engineering analysis of a commercial-scale reprocessing plant prepared by DOE's Savannah River Site found that separation of cesium and strontium could account for 25 percent of the plant's life-cycle cost and over 20 percent of its area and could reduce the plant's performance and reliability because of the engineering challenges involved.

Representatives of two of the industry consortia that received funding under GNEP have expressed similar concerns about separating cesium and strontium and have instead suggested alternatives. For example, one suggestion is to keep cesium and strontium with other high-level radioactive waste and store the waste temporarily, for decades rather than centuries, to allow some radioactive decay before disposal in a geologic repository. Such alternatives may not achieve the same extension of Yucca Mountain's capacity estimated by Argonne National Laboratory but nevertheless indicate the potential insights DOE can attain by working with industry. DOE officials told us they agree that working with industry is critical under either its original approach for an engineering-scale demonstration or its accelerated approach of building commercial-scale facilities, and DOE is considering industry suggestions for alternatives to separating cesium and strontium.

DOE's Original Approach to GNEP Included Building a Separate Engineering-Scale Reprocessing Plant before Conducting R&D that Would Help in Designing the Plant

DOE's original schedule for building the three facilities envisioned under GNEP called for an engineering-scale reprocessing plant to start up between 2011 and 2015—several years before the R&D facility and the fast reactor, which would start up between 2014 and 2019. The more recent GNEP technology development plan pushed back the schedule for all three facilities, with the reprocessing plant starting up around 2020, the

R&D facility between 2020 and 2022, and the fast reactor between 2022 and 2024. Regardless of the precise dates, scheduling the engineering-scale reprocessing plant before the other two facilities would not allow testing and development conducted at the other two facilities, particularly the R&D facility, to be incorporated into the design of the plant.

Specifically, the reprocessing plant would not benefit from testing and development on recycled fuel and advanced reprocessing and safeguards technologies. The recycled fuel R&D schedule spans about 20 years, beginning with testing small samples of different types of recycled fuel and progressing to entire fuel assemblies, which would be fabricated in the R&D facility and tested in the fast reactor. DOE is at the beginning of this effort and has not yet developed technology to overcome key challenges, such as how to remotely fabricate highly radioactive recycled fuel. Given the 20-year fuel development schedule, an engineering-scale reprocessing plant built before making further progress on fuel R&D would increase the risk that the plant would separate transuranics in a form not suitable for fabrication into the type of recycled fuel DOE ultimately chooses to develop. The DOE Savannah River Site's engineering analysis of a commercial-scale reprocessing plant ranked the risk of incompatibility between the output of the plant's spent fuel separations process and recycled fuel fabrication as the most severe programmatic risk associated with the plant. In addition, an engineering-scale reprocessing plant built before the R&D facility could not initially take advantage of advanced reprocessing and safeguards technologies that DOE intends to test and develop at the R&D facility. While DOE national laboratories are currently conducting R&D on such technologies at existing facilities, the testing is generally at a smaller scale, using kilogram quantities of spent fuel, than would be possible at the R&D facility envisioned under GNEP, which would be designed to handle metric tons of spent fuel.

Under DOE's original time frame, an engineering-scale reprocessing plant would also be built earlier than needed because it would separate transuranics before the fast reactor would recycle them as fuel. DOE's plans for the fast reactor do not call for it to initially use recycled fuel produced by the reprocessing plant. It would instead start up using conventional fast reactor fuel, consisting of either uranium or acombination of uranium and plutonium. Recycled fuel assemblies, which would initially be fabricated at the R&D facility, would only gradually begin to replace the conventional fuel as R&D on the recycled fuel nears completion. Thus, from the standpoint of providing sufficient quantities of recycled fuel for the first fast reactor, the reprocessing plant would not be needed until the reactor's need for recycled fuel exceeded the fabrication capacity of the R&D facility.

While a separate engineering-scale reprocessing plant would not initially be needed, it could serve at a later point to increase the maturity of advanced recycling technologies prior to commercialization and demonstrate the technologies in an industrial setting with higher requirements for operational efficiency and continuity of operations than an R&D facility. An expert panel convened by DOE recommended an annual throughput of 100 metric tons of spent fuel as sufficiently large to demonstrate the feasibility of scaling up to a commercial plant, which could have an annual throughput of as much as 2,000 to 3,000 metric tons. Jumping directly from an R&D facility to a commercial-scale reprocessing plant would increase the risk that new technologies would not work as intended. In fact, the Savannah River Site engineering analysis of a commercial-scale reprocessing plant placed a high risk on the possibility that a plant using new processes would require changes or adjustments during or following startup and stated that unanticipated problems requiring equipment modification or replacement would be likely. A recent report by the National Academies echoed this concern and recommended engineering-scale facilities for GNEP because they could be modified faster and at less cost than large-scale facilities. An engineering-scale reprocessing plant would also cost substantially less to build than a commercial-scale plant. DOE's March 2006 mission need statement for GNEP estimated the cost of an engineering-scale plant at between $0.7 billion and $1.7 billion. In contrast, DOE has suggested that the cost of a commercial plant could be estimated by scaling up the almost $20 billion cost of an 800-metric ton reprocessing plant built in Japan. Using this approach, and DOE's guideline for scaling facilities of different sizes, a reprocessing plant with an annual throughput of 3,000 metric tons of spent fuel per year could cost approximately $44 billion. The Savannah River Site's engineering analysis of a 3,000-metric ton reprocessing plant suggests that the cost could also be significantly higher than $44 billion given the uncertainties in designing a plant to use new technologies.

An alternative to building a new engineering-scale reprocessing plant is to modify an existing facility at a DOE national laboratory; however, this alternative may not be cost-effective. The Savannah River Site studied the feasibility of modifying two existing DOE facilities that are not currently being used—the F Canyon at the Savannah River Site and the Fuel Processing Restoration facility at Idaho National Laboratory. The study found that, while the facilities would be capable of supporting an engineering-scale demonstration, both would require major modifications because they are contaminated from previous use or were designed for other purposes. The study estimated the cost to backfit the facilities at $1.3 billion to $1.9 billion and $5.4 billion to $7.9 billion, respectively.

The R&D Facility and Advanced Reactor Would Enable DOE to Develop the Advanced Recycling Technologies Envisioned under Its Original Approach to GNEP

Under DOE's original approach to GNEP, the R&D facility and fast reactor would enable the DOE national laboratories to increase the maturity of advanced recycling technologies and to conduct the required R&D that existing DOE facilities have limited capability to support. Many of the advanced recycling technologies that were the focus of DOE's original approach to GNEP are at a low level of maturity and would benefit from such R&D. For example, testing of DOE's advanced reprocessing technologies has to date been conducted at the laboratory scale, using at most kilogram quantities of spent fuel and with discrete reprocessing steps performed separately rather than continuously, as in a commercial plant. (See app. II for more information on the method DOE has used to assess the maturity of spent fuel recycling technologies and the results of its assessment.) Under its original approach, DOE estimated the cost of the R&D facility at $1.5 billion to $3 billion and the cost of the initial fast reactor at $2 billion to $5 billion.

The R&D facility would provide capabilities—particularly testing and development of recycled fuel and advanced reprocessing and safeguards technologies—that the DOE laboratories currently lack. DOE's plan for developing recycled fuel containing transuranics calls for the R&D facility to develop remote fabrication techniques for the fuel and to actually fabricate recycled fuel assemblies for testing in a fast reactor. While existing facilities at DOE national laboratories can fabricate start-up fuel for the fast reactor, they have limited capability to fabricate transuranic-bearing recycled fuel, which would be more radioactive than start-up fuel and require specialized facilities with heavy shielding to protect workers. DOE also plans for the R&D facility to have a high level of flexibility and range of capabilities so that it can help resolve technical challenges associated with advanced reprocessing technologies. A further advantage of the R&D facility is that it would enable reprocessing R&D to be integrated with fuel fabrication, thereby minimizing shipments of radioactive materials among national laboratories. A fast reactor, like the R&D facility, would also provide capabilities that DOE currently lacks. Inparticular, while DOE can test small samples of recycled fuel either in domestic facilities that approximate conditions in a fast reactor or in fast reactors operated in other countries, a fast reactor built and operated in the United States would enable DOE to test full-scale recycled fuel assemblies. Testing of full-scale assemblies would be required to demonstrate safety and obtain approval by NRC, which would, in turn, enable the commercialization and construction of additional fast reactors capable of using recycled fuel.

A decision to proceed with design and construction of an R&D facility and fast reactor would present DOE with choices regarding the size of the facilities and whether to rely on existing facilities as an alternative to new ones. Existing DOE national laboratory facilities large enough for laboratory-scale R&D on advanced reprocessing technologies have limitations, and some require upgrades. For example, Argonne National Laboratory cut back R&D on advanced reprocessing technologies after the laboratory director decided in October 2007 not to pursue necessary safety upgrades at a facility due to lack of funding. Argonne instead transferred the R&D to another laboratory. Despite such limitations, DOE is evaluating the cost and benefits of using existing laboratory facilities as an alternative to building all or part of a new R&D facility.

Design and construction of a fast reactor would also present choices. DOE does not currently have plans to restart the last one to operate, the Fast Flux Test Facility in Washington state, which is currently being deactivated pending decommissioning. DOE officials believe the cost to restart the facility could be in excess of $500 million.[2] While it could be used to test full-scale fuel assemblies, DOE officials noted that the facility is not well-suited for demonstrating innovative technologies for cost reduction and competitive electricity generation, which would be needed for future commercialization of fast reactors. In terms of building a new fast reactor, DOE is evaluating a wide range of sizes. Under DOE's original approach to GNEP, Argonne National Laboratory, the lead laboratory for fast reactor development, evaluated sizes for the initial reactor ranging from 125 to 840 megawatts.[3] The laboratory concluded that 250 megawatts would balance the need for a realistic test environment against the increased complexity and construction cost of a larger reactor. However, according to the DOE official in charge of fast reactor development, a 250 megawatt reactor might not be large enough to demonstrate competitive electricity generation. Thus, DOE is evaluating larger sizes, up to 3,000 megawatts, to determine the size that would best support the reactor's commercialization.

---

[2]The Columbia Basin Consulting Group, which favors restarting the facility, developed the $500 million estimate. DOE officials do not consider the estimate to be reliable because it was developed quickly and has not been independently validated.

[3]These sizes are expressed in thermal power, which is the gross power of a reactor and does not take into account the efficiency of conversion to electricity.

DOE's Accelerated Approach Would Likely Rely on Technologies That Fall Short of Meeting GNEP's Objectives

Two of the four industry consortia that DOE has funded under its accelerated approach to GNEP have proposed using unproven evolutions of current technologies—particularly the recycling of MOX fuel in existing reactors—that would reduce waste and mitigate proliferation risks to a much lesser degree than anticipated from the advanced technologies envisioned under DOE's original approach. In contrast, the other two consortia proposed technologies that would address GNEP's waste reduction and nonproliferation objectives; however, the technologies are not mature enough for commercial deployment and would therefore not allow DOE to accelerate design and construction of commercial-scale facilities. Under any of the proposals, DOE is unlikely to meet its goal of deploying the facilities in a way that will not require a large amount of government funding. DOE officials recognize these limitations and instead point to other benefits of its accelerated approach.

Two Industry Consortia Have Proposed Using Evolutions of Current Technologies for Addressing GNEP's Objectives

Two of the four industry consortia that received funding have submitted proposals for using unproven evolutions of current recycling technologies that would represent at best an intermediate step toward meeting GNEP's waste reduction and nonproliferation objectives. The proposals call for the initial reprocessing plant to produce MOX fuel (a mixture of plutonium and uranium), or a variant of MOX, for use in existing reactors—a technology choice that would not sufficiently reduce the quantity of transuranics in the high-level radioactive waste stream to meet GNEP's waste reduction objective. The two industry consortia also made proposals for dealing with the transuranics not recycled as part of the MOX fuel in existing reactors. However, the proposals rely on advanced technologies that are at a low level of maturity and would require substantial R&D; implementation of such technologies at a commercial scale would very likely need to follow after implementation of MOX technologies. Although DOE officials involved in managing GNEP have recently expressed support for MOX technologies, the January 2007 GNEP strategic plan rules out MOX on the grounds that it would offer a minor benefit to a geologic repository but not meet GNEP's objectives.

According to DOE estimates, using MOX fuel could increase by about 10 percent the amount of waste that could be disposed of in a geologic repository limited by temperature considerations. In contrast, as discussed earlier, Argonne National Laboratory has estimated that successful development of advanced technologies for recycling transuranics could increase such a repository's capacity almost sixfold, or by almost 600 percent.

DOE officials said that, given the minor waste benefit associated with MOX technologies, they would only pursue MOX technologies as part of a plan to continue to develop more advanced technologies. Specifically, DOE and others have concluded that fast reactors are critical to the ability to recycle transuranics. Even in countries such as France, which currently operates recycling facilities that produce MOX fuel for light water reactors, development of fast reactors that use transuranics is a long-term goal. According to Electric Power Research Institute staff, France did not originally intend for its reprocessing plant to produce MOX fuel for light water reactors; rather, it developed MOX programs because fast reactor technology did not progress as planned and the country needed to address the costs associated with interim storage and safeguarding of plutonium that had been separated out through reprocessing.

Both industry proposals for using evolutions of current recycling technologies to produce MOX fuel would also require DOE to accept less proliferation-resistant technologies than the department envisioned when MOX was not under consideration as part of GNEP. DOE's National Nuclear Security Administration has raised proliferation concerns about MOX technology, particularly MOX fuel fabrication, and indicated in a May 2006 GNEP program document that phasing out current reprocessing technologies (i.e., PUREX) and civilian MOX programs worldwide would provide nonproliferation benefits. While the evolutionary technologies would offer some improvement over existing MOX technologies because they would not separate out pure plutonium, the plutonium mixtures proposed for recycling into MOX fuel would be less proliferation resistant than the mixture produced under DOE's original preferred option (UREX+1a), which would keep plutonium mixed with additional transuranics. For example, pure plutonium could be obtained from a plutonium-uranium mixture for producing MOX fuel without using any heavy shielding from radiation. Moreover, because DOE's schedule for the reprocessing plant calls for it to begin operation at roughly the same time as the proposed R&D facility, the plant would not incorporate advanced nonproliferation safeguards that the R&D facility would develop. (As discussed earlier, the engineering-scale reprocessing plant envisioned under DOE's original approach would also face this limitation, as would any reprocessing plant designed and built prior to the R&D facility.) Instead, DOE officials have suggested that any new reprocessing plant built in the United States would incorporate the latest safeguards technologies available and would also be designed to accommodate more advanced safeguards as they are developed.

The proposed evolutionary technologies build upon existing commercial technologies but are in some respects unproven, and their first deployment at a commercial scale would likely be in the GNEP facilities. For example, one of the consortia proposed using a process for keeping plutonium mixed with uranium that, according to the GNEP separations technologies campaign manager, has only been validated at a laboratory scale. DOE's ability to meet its nonproliferation objective would be reduced if the technologies were not successfully developed and DOE fell back on less advanced technologies for producing MOX fuel, as both industry consortia have proposed as a backup option. In particular, such a backup option could result in a reprocessing plant that separates out plutonium, as is done in Japan. Other technologies that would likely be demonstrated for the first time at a commercial scale in GNEP facilities include technologies for controlling certain radioactive emissions from the reprocessing plant, which would be needed to meet U.S. environmental regulations.

If these unproven technologies associated with producing MOX fuel for existing nuclear power plants can be successfully developed, they could allow the United States to begin recycling spent fuel sooner and on a larger scale than if DOE relied on more advanced but less mature technologies. Specifically, the two industry consortia proposed building a plant by 2023 that could reprocess from 800 to 1,500 metric tons of spent fuel per year. This throughput would be closer to the rate at which existing nuclear power plants produce spent fuel—about 2,200 metric tons per year—than the throughput of an engineering-scale plant.

Two Other Industry Consortia Proposed to Address GNEP's Objectives by Using Technologies That Are Not Mature Enough for Commercial Deployment

DOE would not be able to accelerate deployment of commercial-scale facilities using technologies proposed by the remaining two industry consortia that received DOE funding. As explained in the GNEP strategic plan, one of DOE's goals in working with industry is to avoid the need for engineering-scale facilities and to increase the speed of arriving at a commercially operated system of prototype recycling facilities. However, the two consortia proposed technologies that are in some cases no more mature or even less mature than the advanced technologies DOE had planned to demonstrate under its original approach. For example, one industry consortium proposed to rely on the type of reprocessing technology (UREX+) that DOE has been developing, which the department had planned to demonstrate at an engineering scale. The consortium also proposed a "two-tier" system in which transuranics would first be recycled in an advanced thermal reactor,[14] then in a fast reactor. However, the advanced thermal reactor technology is still being developed, and implementing a two-tier system with dual sets of technologies would significantly increase the need for R&D. The other industry consortium proposed a type of advanced reprocessing technology (electrochemical) that DOE considers even less mature for reprocessing light water reactor spent fuel than the UREX+ technologies being developed by DOE. Thus, under either industry proposal, skipping the engineering phase of development would create an undue risk that the technologies would not work as intended.

On the other hand, the consortia's proposed technologies would, if successfully developed, address GNEP's waste and nonproliferation objectives by recycling transuranics in fast reactors and keeping plutonium mixed with the other transuranics. Representatives of both consortia have also argued that some of their proposed technologies are superior to DOE's—for example, that electrochemical reprocessing would provide a greater intrinsic barrier to proliferation than DOE's technologies, in part because spent fuel would be processed in batches, thereby facilitating efforts to track the materials separated from spent fuel and to detect theft or diversion. Similarly, representatives of the consortium proposing the two-tier system stated that their proposed combination of technologies would reduce energy costs compared with recycling only in fast reactors—for example, because an advanced thermal reactor would extract more energy from recycled fuel and convert the energy more efficiently to electricity than a fast reactor.

The Government Would Likely Bear Substantial Costs for Commercial-Scale Recycling Facilities

DOE has cited industry's potential willingness to invest substantial sums of private money to construct and operate GNEP facilities as a reason for considering commercial-scale facilities. Furthermore, the GNEP strategic plan established a goal of developing and implementing such facilities in a way that will not require a large amount of government construction and operating funding to sustain. However, our review of industry proposals and interviews with DOE officials indicate that the department is unlikely to meet this goal, at least for the first GNEP facilities. Some industry proposals state, for example, that initial facilities would rely entirely on government support and that the need for such support would be reduced only after demonstration of new recycling technologies in the initial facilities and development of cost-saving features.

Most notably, industry has generally proposed that design and construction of the initial fast reactor be funded directly by DOE, perhaps with ongoing government funding or other incentives, such as fees paid to the reactor operator for using recycled fuel. According to DOE, the industry proposals estimated the cost of the initial fast reactor at $2 billion to $4.5 billion—a cost that may be understated given that DOE's estimate for the cost of a smaller test reactor under its original approach to GNEP was roughly the same: $2 billion to $5 billion. DOE funding would be required because fast reactors are initially expected to be more expensive to build and operate than light water reactors and thus unable to compete with them economically based on sales of electricity alone. According to DOE, studies by the Nuclear Energy Agency have estimated that a fast reactor's capital costs, for example, may be about 25 percent higher than those for light water reactors. Furthermore, components that would help reduce the cost of a fast reactor and make it more economically competitive are at a relatively low level of maturity and, according to some industry responses, would not be ready for commercial-scale deployment by DOE's time frame of 2025.

DOE officials recognize that industry will not pay for design and construction of the initial fast reactor and have considered two other options: delaying the reactor or sharing the cost with other countries that are also interested in developing fast reactors. According to the DOE official in charge of fast reactor development, delaying the reactor is a possibility if the department decides in favor of recycling MOX fuel in light water reactors. In addition, DOE has negotiated a memorandum of understanding with Japan and France to harmonize fast reactor development efforts. DOE officials have expressed hope that Japan and France would contribute to the cost of building a fast reactor in the United States, where it could be licensed by NRC. A reactor with NRC approval would, in turn, have a greater potential for commercialization because utilities would have a higher degree of confidence in the technology.

For the reprocessing plant, two of the industry consortia again proposed direct funding by DOE. Other funding possibilities might reduce the government's financial burden but could still require significant government support. For example, according to DOE, revenues from sales of MOX fuel produced by the plant for use in light water reactors would not significantly offset the plant's capital costs and would not attract sufficient private investment. To use MOX fuel, U.S. reactors would typically have to undergo some physical modification and receive a license amendment from NRC. Thus, even though it is generally more expensive to produce, MOX fuel may have to be sold at a discount compared with conventional fuel in order for it to be commercially attractive to U.S. utilities. For example, to ensure a market for MOX fuel that is to be produced at a DOE facility for recycling surplus weapons-grade plutonium, DOE has agreed to provide the MOX fuel at a discounted price and to pay for the necessary modifications to light water reactors where it will be used. Another funding option proposed by industry is obtaining private financing backed by federal loan guarantees or federal contracts to treat a specified volume of spent nuclear fuel at a set price that would cover operating the plant and servicing the debt. The government could incur a liability under such options if industry defaulted on loans or, depending on the specific conditions of such funding arrangements, if factors such as litigation or regulatory delays prevented the plant from reprocessing spent fuel.

Industry has also proposed, in expressions of interest and deliverables submitted to DOE, that at least a portion of the fee that nuclear power plant operators now pay into the Nuclear Waste Fund—a special fund under DOE's jurisdiction, subject to annual appropriations by Congress, for disposal of spent fuel in a geologic repository—be used to pay for a commercial reprocessing plant.[1] Proponents of this option have called for establishing a separate

---

[1] The fee is currently set at one mil ($0.001) per kilowatt-hour of electricity generated and sold by the power plants.

government entity that could access the fund, potentially without the need for annual appropriations. They have also suggested that the current fee be increased to cover the full costs of spent fuel disposal, including the cost of both a geologic repository and a reprocessing plant. Implementing this proposal would require substantial legislative and regulatory changes. For example, the Nuclear Waste Policy Act does not allow the Nuclear Waste Fund to be used for reprocessing activities. DOE officials said that, while they recognize that current legislation limits how the Nuclear Waste Fund can be used, they would not rule out proposals to use the fund for GNEP. Instead, a decision by the Secretary of Energy to support such proposals would be contingent on a change in legislation.

In addition to requiring direct government funding, working with industry to design and build commercial-scale facilities would also likely require that DOE invest significant R&D resources. DOE national laboratories would need to conduct some of the R&D even under DOE's original plan for an engineering-scale demonstration—for example, on technology for capturing radioactive emissions from a reprocessing plant. However, industry has requested DOE assistance with other R&D, such as MOX fuel certification, that could divert resources from advanced technologies ultimately needed to meet GNEP's objectives. According to the head of GNEP's technical integration office, the national laboratories would give long-term R&D on advanced technologies a lower priority than industry's immediate R&D needs.

DOE Officials Recognize the Limitations of Accelerating Deployment of Commercial-Scale Facilities but Cite Other Benefits

Given the technologies industry can provide in DOE's time frame, an accelerated approach would likely require the department to recycle MOX fuel in existing commercial nuclear power plants. DOE officials acknowledge that MOX recycling technologies would constitute an intermediate step toward GNEP's objective of reducing radioactive waste and that achieving this objective would ultimately require development of advanced technologies for recycling transuranics in fast reactors. Nevertheless, according to the officials, working with industry to deploy commercial-scale facilities to recycle MOX fuel in existing reactors would provide enough of a waste reduction benefit to allow time to develop more advanced technologies. The proposal to build GNEP facilities for recycling MOX fuel in existing reactors as an intermediate step is similar to a plan

DOE put forward in 2005 under the Advanced Fuel Cycle Initiative, prior to the announcement of GNEP in February 2006.[2] The plan called for developing the ability to recycle spent nuclear fuel in evolutionary stages, with each stage helping to develop technology required for the next and providing successively greater benefits in terms of extending the technical capacity of a geologic repository. According to DOE, the department is evaluating the possibility of revising the GNEP strategic plan to allow for the possibility of recycling MOX fuel in existing reactors, as previously contemplated under the Advanced Fuel Cycle Initiative.

With regard to nonproliferation, DOE officials emphasize that the international benefits of working with industry to deploy commercial-scale facilities outweigh what DOE considers to be the manageable risk of nuclear material theft from such a facility built domestically. In particular, DOE officials consider deploying commercial-scale recycling facilities as essential for the United States to play a leadership role among countries with advanced nuclear capabilities and to persuade other countries that they should rely on international fuel services rather than developing domestic uranium enrichment or spent fuel reprocessing capabilities. While they have not ruled out other industry proposals, DOE officials have also cited nonproliferation benefits of recycling MOX fuel in light water reactors, such as the ability to reduce stocks of plutonium that accumulate in spent nuclear fuel; reducing and eventually eliminating excess stocks of civilian plutonium is part of the nonproliferation objective set forth in the GNEP strategic plan. DOE's Nuclear Energy Research Advisory Committee has indicated that it may be appropriate to consider using existing reactors for this purpose, particularly if large-scale deployment of fast reactors, which would also be capable of reducing plutonium stocks, does not occur until the middle of the century.

Finally, DOE has argued that the government's investment in commercial-scale spent nuclear fuel recycling facilities would be worthwhile. DOE officials have said that the benefit of ensuring U.S. leadership on nonproliferation issues through the construction of commercial-scale facilities—even ones that rely on evolutionary MOX technologies—would outweigh the cost to the government. Furthermore, the officials have suggested that revenues generated by the facilities, such as through the sale of MOX fuel, would at least offset some of the government's cost. Utilities that operate commercial nuclear power plants might be interested in MOX fuel because it could provide an alternative to uranium fuel, supplies of which could become limited given the worldwide growth of nuclear

---

[2] For more information on this plan, see DOE, Advanced Fuel Cycle Initiative: Objectives, Approach, and Technology Summary (Washington, D.C., May 2005).

energy. Over the longer term, DOE has argued that recycling spent nuclear fuel would be an attractive option to the government if the cost of doing so were comparable to direct disposal, which could require design and construction of multiple geologic repositories. If DOE chooses to rely on MOX technologies, this argument hinges on a later transition to more advanced technologies since, as discussed earlier, recycling MOX fuel in existing reactors would provide a minor waste reduction benefit.

CONCLUSIONS

Accelerating deployment of commercial-scale spent nuclear fuel recycling facilities that have a limited impact on GNEP's waste reduction and nonproliferation objectives would take DOE down a costly path that would likely draw resources away from developing the advanced technologies ultimately needed to meet these objectives. The technologies closest to being commercially available are evolutions of existing MOX fuel recycling technologies that would reduce waste and mitigate proliferation risks to a much lesser degree than is anticipated from advanced technologies for recycling all of the transuranics in spent nuclear fuel. If DOE pursues an accelerated approach to deploying commercial-scale facilities, the timing of a transition to more advanced technologies that fully meet GNEP's waste reduction and nonproliferation objectives is unclear because such technologies are at a low-level of maturity and require significant R&D. Accelerating deployment of commercial-scale facilities could serve as an intermediate step—but a costly one. While the GNEP strategic plan suggests that such facilities would need little government financial support, industry proposals suggest the opposite. As a result, the level of government financial commitment needed to deploy commercial-scale facilities would likely draw resources away from R&D on more advanced technologies and create a risk of delaying rather than accelerating progress toward ultimately meeting GNEP's waste reduction and nonproliferation objectives.

DOE's original approach of demonstrating advanced technologies at an engineering scale appears more likely over the long term to address GNEP's waste reduction and nonproliferation objectives than the department's accelerated approach. Nevertheless, an engineering-scale demonstration is not without risks, including the possibility that advanced recycling technologies currently at a low level of maturity might not perform as expected and might not be commercially viable. DOE's original approach to GNEP in some respects increased these risks. In particular, an engineering-scale reprocessing plant built according to DOE's original schedule—before an R&D facility and advanced reactor that would support testing and development of recycled fuel—could result in a plant that separates the materials in spent fuel in a form unsuitable for recycled fuel fabrication. The schedule would also not allow the plant to incorporate advanced safeguards and reprocessing technologies developed at the R&D facility. With regard to commercial viability, DOE's engineering-scale approach lacked industry participation that could help promote future commercialization and widespread use of the advanced technologies. DOE's efforts to work with industry under its accelerated approach to GNEP have mitigated some of the risk that DOE might focus on developing overly costly and complex technologies, and working with industry under its engineering-scale approach could continue to mitigate this risk.

RECOMMENDATIONS FOR EXECUTIVE ACTION

We recommend that the Secretary of Energy direct the Office of Nuclear Energy to reassess its preference for an accelerated approach to implementing GNEP through construction of commercial-scale facilities using spent nuclear fuel recycling technologies that industry can offer in DOE's time frame. The reassessment should consider the time and government resources required to support both the initial spent nuclear fuel recycling facilities and R&D on more advanced recycling technologies that fully meet GNEP's objectives.

If DOE decides to pursue design and construction of engineering-scale facilities for demonstrating advanced technologies, we further recommend that the Secretary of Energy take the following two actions:

- Revise the schedule for an engineering-scale reprocessing plant so that the plant is built after an R&D facility and advanced reactor have conducted sufficient testing and development of recycled fuel to ensure that the output of the reprocessing plant can be fabricated into recycled fuel and used in an advanced reactor. The revised schedule should also allow for the R&D facility to test and demonstrate advanced reprocessing and safeguards technologies that would be used in the reprocessing plant.
- Direct the Office of Nuclear Energy to work with industry to the extent possible on advanced spent nuclear fuel recycling technologies in order to obtain industry's expertise and input on future commercialization of such technologies.

Agency Comments and Our Evaluation

We provided a draft of this report to DOE and NRC for their review and comment. DOE's written comments are reproduced in appendix III. DOE agreed with many of our findings and concurred with our recommendations, directed toward the department's original engineering-scale approach to GNEP, to revise its schedule for an engineering-scale reprocessing plant and to work with industry to the extent possible. With regard to our recommendation that DOE reassess its preference for an accelerated approach to implementing GNEP, DOE stated that the department will continue to perform analyses to support the Secretary of Energy's decision on the direction for GNEP. DOE and NRC also provided detailed technical comments, which we have incorporated into our report as appropriate.

DOE raised several issues with our draft report. First, DOE stated that the report gives an erroneous impression that fast reactors can never be economically competitive with light water reactors. We have clarified the report to indicate that fast reactors are at least initially expected to be more expensive to build and operate than light water reactors. We recognize that one of DOE's research goals is to develop fast reactors that are competitive with light water reactors. However, as noted in our report, technologies that would help make fast reactors more economically competitive are at a low level of maturity. The low level of maturity of such technologies is a key reason that industry has proposed the first fast reactor envisioned under GNEP be funded by DOE.

Second, DOE stated that the report gives an erroneous impression that recycling MOX fuel in light water reactors in the near-term would have a limited impact on GNEP's waste reduction and nonproliferation objectives and would draw resources away from developing advanced technologies in the long term.

We disagree. With regard to waste reduction, our report accurately states that the GNEP strategic plan specifically rules out using MOX in light water reactors because it would offer a minor waste reduction benefit but not meet GNEP's objectives. Now that the department is considering evolutionary MOX technologies, DOE cited the substantial reduction in the quantity of spent nuclear fuel in storage as a significant near-term benefit of recycling in light water reactors. Our report acknowledges that such a MOX program could allow DOE to begin recycling spent fuel sooner and on a larger scale than more advanced but less mature technologies. Furthermore, we have clarified the report to show that DOE has indicated it would only pursue evolutionary MOX technologies as part of a plan to later transition to more advanced technologies for recycling in fast reactors, which are anticipated to provide a much greater waste reduction benefit than evolutionary MOX technologies from the standpoint of extending the capacity of a geologic repository. The question, in our view, is whether the intermediate benefit of reducing the quantity of spent nuclear fuel in storage would be worth the investment in evolutionary MOX technologies. On this point, DOE stated that facilities for recycling spent fuel in light water reactors would be funded and constructed by industry only when justified by a sound business case, without impacting government funding for R&D on more advanced recycling technologies. In contrast, our report points out that industry does not expect the evolutionary MOX technologies to be profitable—at least under current conditions—without some form of government support and R&D assistance. Thus, while it is conceivable that the government could provide the necessary support and R&D assistance while also continuing to fund R&D on more advanced technologies, the evolutionary technologies could also draw resources away from the more advanced technologies.

With regard to nonproliferation, DOE called into question our finding that evolutionary MOX technologies would mitigate proliferation risks to a lesser degree than anticipated from the advanced technologies envisioned under the engineering-scale approach to GNEP. Rather than differentiating between the proliferation resistance of alternative reprocessing technologies, DOE stated that any reprocessing plant, if misused, could be modified to create weapons usable material. Thus, it is the department's view that its nonproliferation objectives would be largely accomplished through international policies that seek to avoid the spread of enrichment and reprocessing technologies while eliminating existing plutonium inventories and production of material mixes that are attractive for use in creating a nuclear explosive. We recognize that the degree of proliferation resistance of reprocessing technologies is only one aspect of GNEP's nonproliferation objective. Nonetheless, our report is consistent with the GNEP technology development plan, which states that the reprocessing technology preferred under the original approach to GNEP (UREX+1a) provides an additional degree of proliferation resistance compared with other processes precisely because it would not separate plutonium from any of the transuranics. Based on this reasoning, UREX+1a would also provide an additional degree of proliferation resistance compared with evolutionary MOX technologies that, for example, keep plutonium mixed with uranium but not with other transuranics.

As we agreed with your offices, unless you publicly announce the contents of this report earlier, we plan no further distribution until 30 days from the date of this letter. At that time, we will send copies of this report to interested congressional committees, the Secretary of Energy, the Chairman of the Nuclear Regulatory Commission, and other interested parties. We will also make copies available to others upon request. In addition, the report will be available at no charge on the GAO Web site at http://www.gao.gov.

If you or your staffs have any questions about this report, please contact me at (202) 512-3841 or aloisee@gao.gov. Contact points for our Offices of Congressional Relations and Public Affairs may be found on the last page of this report. Other staff contributing to this report are listed in appendix IV.

Gene Aloise
Director, Natural Resources and Environment

List of Committees

The Honorable Carl Levin
Chairman
The Honorable Norm Coleman
Ranking Member
Permanent Subcommittee on Investigations
Committee on Homeland Security and Governmental Affairs
United States Senate

The Honorable Jeff Bingaman
Chairman
Committee on Energy and Natural Resources
United States Senate

The Honorable John D. Dingell
Chairman
The Honorable Joe Barton
Ranking Member
Committee on Energy and Commerce
House of Representatives

The Honorable Bart Gordon
Chairman
The Honorable Ralph M. Hall
Ranking Member
Committee on Science and Technology
House of Representatives

The Honorable Edward J. Markey
Chairman
Select Committee on Energy Independence and Global Warming
House of Representatives

The Honorable Bart Stupak
Chairman
The Honorable John Shimkus
Ranking Member
Subcommittee on Oversight and Investigations
Committee on Energy and Commerce
House of Representatives

APPENDIX I.
SCOPE AND METHODOLOGY

To evaluate the Department of Energy's (DOE) original engineering-scale approach to implementing the Global Nuclear Energy Partnership (GNEP), we analyzed (1) how DOE had selected the advanced spent nuclear fuel recycling technologies on which to focus its research and development (R&D), (2) the department's assessment of the maturity of those technologies, and (3) the plan for developing them:

- We analyzed DOE's selection of advanced technologies by reviewing the department's annual Advanced Fuel Cycle Initiative comparison reports, which assess alternative recycling technologies against waste reduction, nonproliferation, and other criteria. We also reviewed related DOE national laboratory documents, including technical analyses of recycling technologies not selected for development under GNEP. We compared DOE's selection with assessments conducted by independent organizations and entities with expertise in recycling of spent nuclear fuel, including the Nuclear Energy Agency of the Organisation for Economic Co-operation and Development and DOE's Nuclear Energy Research Advisory Committee, and the National Research Council of the National Academies.[1] We interviewed officials of the DOE Office of

---

[1] Two key National Research Council reports we reviewed include National Academy Press, Review of DOE's Nuclear Energy Research and Development Program (Washington, D.C., Oct. 29, 2007), and Nuclear Wastes: Technologies for Separations and Transmutation (Washington, D.C., 1996).

Nuclear Energy, the National Nuclear Security Administration, and DOE national laboratories regarding the selection of advanced technologies under GNEP.
- We analyzed DOE's assessment of the maturity of advanced recycling technologies as presented in the GNEP technology development plan. We specifically analyzed how DOE had used technology readiness levels (TRL), a method for ranking the maturity of technologies, and compared DOE's use of the method to Department of Defense guidance for technology readiness assessments. We also interviewed DOE and DOE national laboratory officials about the maturity of the technologies and their use of TRLs. We observed R&D activities related to development of advanced reprocessing, fast reactor, waste form, and recycled fuel technologies at four DOE national laboratories (Argonne, Idaho, Los Alamos, and Oak Ridge) and interviewed DOE national laboratory researchers about their efforts. We selected the laboratories based on their leading roles in implementing spent fuel recycling R&D. We also observed facilities used for R&D on safeguards technologies at Idaho State University's accelerator center, which we elected to visit because of its proximity to Idaho National Laboratory.
- We analyzed DOE's plan for developing advanced spent nuclear fuel recycling technologies as presented in the GNEP technology development plan, spent nuclear fuel recycling program plan, and mission need statement; DOE's budget justifications for the Advanced Fuel Cycle Initiative; and other planning documents. We interviewed DOE officials responsible for managing GNEP, including the officials responsible for directing work on each of the three initial GNEP facilities and for overseeing R&D on advanced recycling technologies. We interviewed DOE national laboratory officials responsible for directing R&D on advanced recycling technologies, including the head of the GNEP technical integration office established by DOE at the Idaho National Laboratory and the seven GNEP campaign managers for systems analysis, separations (i.e., reprocessing), recycled fuel, fast reactors, safeguards, waste forms, and grid-appropriate reactors.[2] We observed DOE national laboratory facilities that DOE has evaluated for use in GNEP as an alternative to building new facilities, particularly the F Canyon at the Savannah River Site and the Fuel Processing Restoration

facility at Idaho National Laboratory. We also interviewed Savannah River Site officials regarding their engineering alternative studies for a commercial-scale reprocessing plant based on the advanced technologies that were the focus of DOE's original approach to GNEP.

To evaluate DOE's accelerated approach of working with industry to design and build commercial-scale recycling facilities, we analyzed DOE documents related to the department's decision to consider working with industry, including the August 2006 request for industry expressions of interest in designing and building a commercial-scale reprocessing plant and fast reactor, the January 2007 GNEP strategic plan, and the funding opportunity announcement for conceptual design studies, business plans, and related documents. Furthermore, we reviewed two sets of documents submitted to DOE: 18 expressions of interest submitted in September 2006 by companies proposing to design and build GNEP facilities and by other entities; and preliminary deliverables submitted in January 2008 by the four industry consortia to which DOE awarded funding pursuant to the funding opportunity announcement. We considered all of these documents, including the less recent expressions of interest, because the terms under which DOE would work with industry are still evolving. Many of the documents contain proprietary information; to protect such information, this report does not disclose details of the various industry responses. We evaluated the documents submitted to DOE to determine the spent nuclear fuel recycling technologies proposed for addressing GNEP's waste reduction and nonproliferation objectives; the maturity of the technologies and the R&D needed to support their use in commercial-scale facilities; and the means proposed for funding initial GNEP facilities. We also reviewed the results of DOE's evaluation of the 18 expressions of interest, as summarized in a November 2006 report, and we interviewed DOE officials regarding their assessment of industry's January 2008 preliminary deliverables. We interviewed representatives of lead firms for the four consortia that received funding under GNEP—AREVA, Energy Solutions, General Electric, and General Atomics—as well as representatives of the Nuclear Energy Institute, which represents the nuclear power industry, and the Electric Power Research Institute.

To evaluate issues of significance to both approaches DOE is considering for implementing GNEP, we interviewed DOE officials in the Office of Nuclear Energy, including the Assistant Secretary for Nuclear Energy (who serves as the GNEP program manager) and the deputy GNEP program manager; the director

---

[2] Development of grid-appropriate reactors scaled for small electricity grids and suited to conditions in developing nations is part of the international component of GNEP and was not a focus of our review.

and other officials of the Office of Civilian Radioactive Waste Management, which is responsible for the Yucca Mountain geologic repository; and the National Nuclear Security Administration, which assists the Office of Nuclear Energy on nonproliferation issues related to GNEP. We also interviewed officials of the Nuclear Regulatory Commission (NRC), which would have regulatory authority over commercial facilities for recycling spent nuclear fuel; and the Nuclear Waste Technical Review Board, an independent agency of the U.S. Federal Government that provides independent scientific and technical oversight of DOE's program for managing and disposing of high-level radioactive waste and spent nuclear fuel. We reviewed DOE's January 2007 notice of intent to prepare a programmatic environmental impact statement for GNEP, and we attended two public hearings on the proposed scope of the programmatic environmental impact statement— one in Ohio near a site being studied to host GNEP facilities and one in Washington, D.C. In addition, we attended DOE's October 2007 annual meeting for GNEP, which included updates on DOE's R&D efforts and plans for initial spent fuel recycling facilities; open meetings related to GNEP convened by NRC's Advisory Committee on Nuclear Waste and Materials and the National Academies; and the American Nuclear Society's 2007 annual meeting, which included sessions related to GNEP and recycling of spent nuclear fuel. Finally, we met with representatives of nongovernmental organizations that have raised concerns about or studied issues related to the implementation of GNEP, such as the Natural Resources Defense Council, the Union of Concerned Scientists, and the Institute for Policy Studies.

We conducted this performance audit from November 2006 to April 2008, in accordance with generally accepted government auditing standards. Those standards require that we plan and perform the audit to obtain sufficient, appropriate evidence to provide a reasonable basis for our findings and conclusions based on our audit objectives. We believe that the evidence obtained provides a reasonable basis for our findings and conclusions based on our audit objectives.

Appendix II.
DOE's Use of Technology Readiness Levels to Assess the Maturity of Spent Fuel Recycling Technologies

The Office of Nuclear Energy has begun to assess the maturity of spent fuel recycling technologies using technology readiness levels (TRL), a method pioneered by the National Aeronautics and Space Administration for measuring and communicating the risks associated with critical technologies in first-of-a-kind applications. The Office of Nuclear Energy also has required that the industry consortia receiving funds under GNEP apply the method to the technologies they propose for deployment. Using a scale from one (basic principles observed) through nine (total system used successfully in project operations), TRLs show the extent to which technologies have been demonstrated to work as intended. Demonstration of new technologies at successively larger scales is one way to increase their maturity, thereby mitigating the risk of cost or schedule overruns in the design and construction of commercial-scale facilities and limiting investment in potentially ineffective technologies. GAO considers seven (subsystem demonstrated in an operational environment) to be an acceptable level of readiness before proceeding with final design and committing to definitive cost and schedule estimates. Based on our review of DOE major projects, we recommended that DOE evaluate and consider adopting a disciplined and consistent approach for assessing TRLs.[1] DOE concurred with our recommendation and has piloted the TRL method in an Office of Environmental Management project, but the department has not decided whether to incorporate the method into its project management guidance.

The Office of Nuclear Energy has adopted the use of TRLs to assess the maturity of spent fuel recycling technologies even though doing so is not a requirement of DOE's project management guidance. The GNEP technology development plan grouped the nine-point scale into three categories: concept development (1 to 3), proof-of-principle (4 to 6), and proof-of-performance (7 to 9). The plan placed virtually all of the advanced spent fuel recycling technologies in the proof-of-principle category: reprocessing of spent fuel produced by both light water reactors and fast reactors; development of new waste forms, which would need to be incorporated into a reprocessing plant to ensure the safe disposal of radioactive waste separated from spent fuel; recycled fuel containing plutonium and other transuranics, in terms of both fabrication and performance; and technologies for reducing the cost of fast reactors.

Based on our review of the technology development plan and interviews with DOE national laboratory officials, some of the advanced technologies are in fact at an even lower level of maturity than indicated in the plan. In particular, the campaign manager for reprocessing technologies provided us with additional information showing that several of the waste forms are at a readiness level of 2 to 3 (concept development) as opposed to 4, as indicated in the plan. Similarly, he provided us with information indicating that some key technologies for reprocessing spent fuel produced by existing light water reactors (i.e., the UREX+ technologies) are at a readiness level of 4 as opposed to 5.

DOE national laboratory officials told us they generally support the use of TRLs to assess the technology maturity and direct limited R&D resources but also pointed out limitations of the method. For example, readiness levels do not indicate the time or resources required to increase the maturity of spent fuel recycling technologies or the obstacles DOE faces. In the case of recycled fuel containing plutonium and other transuranics, the R&D schedule spans about 20 years. DOE is at the beginning of this effort and has already encountered obstacles. For example, DOE so far has not manufactured fuel samples that contain curium, one of the transuranics, because it is highly radioactive and would require remote fabrication techniques that the department has not yet developed.

---

[1] GAO, Department of Energy: Major Construction Projects Need a Consistent Approach for Assessing Technology Readiness to Help Avoid Cost Increases and Delays, GAO-07-336 (Washington, D.C.: Mar. 27, 2007).

Furthermore, DOE plans to rely at least in part on foreign reactors to test the fuel samples, and it was not able to test one of the samples in the French fast reactor where it had planned because of regulatory obstacles in France. The head of the GNEP technical integration office also told us that high readiness levels can mask the challenges DOE would face in designing and building a facility, particularly a fast reactor. The United States has designed and built several fast reactors, so the GNEP technology development plan assigns many of the basic fast reactor components a high readiness level. However, construction on the last fast reactor ended over a quarter century ago. As a result, the United States has lost much of the technical infrastructure and expertise needed to build another reactor.

While the Office of Nuclear Energy deserves credit for adopting the use of the TRLs, despite the method's limitations and the lack of a DOE requirement for using it, we noted areas in which the office could improve its application of the method, particularly if DOE proceeds with its plan to design and build engineering- or commercial-scale recycling facilities. For example, the GNEP technology development plan did not assign TRLs to advanced safeguards technologies even though development of such technologies is important to achieving GNEP's nonproliferation objective.

The campaign manager for safeguards technologies said he had not yet applied the TRL method because the safeguards campaign is new and because existing technologies are adequate for the Nuclear Regulatory Commission (NRC) to license the facilities envisioned under GNEP. Similarly, while the technology development plan assigned TRLs to advanced reprocessing technologies, it did not assign them to the individual separations steps and many pieces of equipment that would make up a reprocessing plant.

INDEX

A

accelerator, 40
access, 27
adaptation, 9
administration, 5
administrative, 5
Advanced Energy Initiative, 5
Advanced Fuel Cycle Initiative, 12, 28, 39, 40
air, 5
air pollution, 5
alternative, 6, 15, 18, 19, 28, 36, 39, 40
alternatives, 1, 12, 16
americium, 8
appendix, 35, 37
application, 45
appropriations, 26
argument, 29
assessment, 10, 19, 39, 40, 41
atoms, 7
auditing, 42
authority, 10, 42
avoidance, 5

B

barrier, 14, 24

barriers, 2
benefits, 2, 20, 21, 23, 28

C

campaigns, 5
cancer, 7
capacity, 1, 6, 7, 8, 9, 10, 12, 13, 15, 16, 17, 22, 28, 36
capital cost, 25, 26
certification, 27
cesium, 8, 15, 16
civilian, 23, 28
Columbia, 20
commercialization, 1, 11, 17, 19, 20, 26, 32, 33
Committee on Homeland Security, 37
complexity, 15, 20
components, 25, 45
confidence, 26
Congress, iv, 26
construction, 3, 19, 20, 21, 25, 28, 33, 43, 45
continuity, 17
contracts, 26
conversion, 20
cost-effective, 18
costs, 22, 25, 26, 27
credit, 45
curium, 8, 44

D

debt, 26
decay, 7, 8, 10, 12, 15, 16
decisions, 10
Department of Defense, 40
Department of Energy (DOE), 7, 39, 44
developing nations, 41
disposition, 8
distribution, 37
draft, 3, 35

E

electricity, 5, 6, 13, 20, 25, 26, 41
energy, 5, 6, 7, 8, 9, 10, 13, 25, 29
environment, 5, 6, 20, 43
environmental impact, 42
environmental regulations, 23
equipment, 14, 18, 45
exclusion, 12
expert, iv, 17
expertise, 33, 39, 45

F

fabricate, 17, 19
fabrication, 17, 19, 22, 32, 44
February, 28
fee, 26
fees, 25
financial support, 31
financing, 26
firms, 41
fission, 7, 8, 15
flexibility, 19
France, 9, 22, 26, 45
fuel, 7, 1, 2, 3, 5, 6, 7, 8, 9, 10, 12, 13, 14, 15, 16, 17, 19, 20, 21, 22, 23, 24, 25, 26, 27, 28, 31, 32, 33, 35, 39, 40, 41, 42, 43, 44, 45
funding, 2, 5, 16, 20, 21, 24, 25, 26, 27, 36, 41
funds, 43

G

GAO, 7, 8, 37, 43, 44
gaseous waste, 10
General Electric, 41
generation, 5, 6, 13, 20
Global Nuclear Energy Partnership (GNEP), i, iii, v, 7, 1, 2, 3, 5, 6, 7, 8, 9, 10, 11, 12, 13, 14, 15, 16, 17, 18, 20, 21, 22, 23, 24, 25, 27, 28, 31, 33, 35, 36, 39, 40, 41, 43, 44, 45
Global Warming, 38
goals, 24, 35
government, iv, 2, 21, 25, 26, 27, 28, 31, 33, 36, 42
greenhouse, 5
greenhouse gas, 5
grids, 41
groups, 7
growth, 28
guidance, 6, 10, 40, 44

H

handling, 14
hazardous materials, 7
heat, 6, 7, 8, 10, 12, 15
high risk, 18
high-level, 2, 6, 8, 12, 15, 16, 22, 42
host, 42
House, 38
humans, 7

I

Idaho, 5, 13, 18, 40
implementation, 5, 12, 22, 42
incentives, 25
inclusion, 9
incompatibility, 17
industrial, 17
industrialized countries, 13
industry, 7, 1, 2, 3, 11, 13, 15, 16, 21, 22, 23, 24, 25, 26, 27, 28, 31, 32, 33, 35, 36, 41, 43
infrastructure, 45

injury, iv
integration, 5, 27, 40, 45
interviews, 2, 25, 44
intrinsic, 24
inventories, 36
Investigations, 37, 38
investment, 7, 2, 26, 28, 36, 43
ions, 5

J

January, 22, 41, 42
Japan, 9, 18, 23, 26
jurisdiction, 26

L

large-scale, 18, 28
lead, 20, 41
leadership, 28
legislation, 27
legislative, 10, 27
licensing, 10
life-cycle, 16
life-cycle cost, 16
limitation, 23
limitations, 2, 19, 21, 44, 45
litigation, 26
loan guarantees, 26
loans, 26
long period, 6
long-term, 1, 9, 11, 22, 27
long-term impact, 1, 11
low-level, 8, 10, 31

M

magnetic, iv
management, 1, 6, 15, 44
market, 26
mask, 45
megawatt, 20
memorandum of understanding, 26
metric, 6, 13, 17, 23

models, 13
money, 25

N

nation, 7, 5, 6, 13
national, 7, 5, 9, 12, 15, 17, 18, 19, 27, 39, 40, 44
National Aeronautics and Space Administration, 43
National Research Council, 39
neptunium, 8
neutrons, 10
New York, iii, iv
nongovernmental, 42
nongovernmental organization, 42
nonproliferation, 2, 5, 21, 22, 23, 24, 28, 31, 35, 36, 39, 41, 42, 45
North America, 5
NRC, 3, 10, 19, 26, 35, 42, 45
nuclear, 7, 1, 5, 6, 7, 9, 10, 12, 13, 14, 23, 26, 27, 28, 31, 33, 36, 40, 41, 42
nuclear energy, 5, 6, 13, 29
Nuclear Energy Agency, 13, 25, 39
nuclear material, 14, 28
nuclear power, 5, 7, 10, 13, 23, 26, 27, 28, 41
nuclear power plant, 5, 10, 13, 23, 26, 27, 28
Nuclear Regulatory Commission, 37, 42, 45
Nuclear Waste Policy Act, 6, 27
nuclear weapons, 9

O

Offices of Congressional Relations and Public Affairs, 37
Ohio, 42
operator, 25
opposition, 16
Organisation for Economic Co-operation and Development, 13, 39
organization, 13
organizations, 12, 39, 42
oversight, 42
oxide, 2, 9

P

Pacific, 5
performance, 6, 16, 42, 44
planning, 40
plants, 5, 10, 13, 24
play, 28
plutonium, 1, 2, 7, 8, 9, 14, 15, 17, 22, 23, 24, 26, 28, 36, 44
pollution, 5
potential energy, 9
power, 5, 7, 10, 13, 20, 23, 26, 27, 28, 41
power plant, 5, 13, 24, 26
power plants, 5, 10, 13, 23, 26, 27, 28
preference, 2, 3, 33, 35
private, 25, 26
private investment, 26
production, 36
program, 13, 22, 36, 40, 41
proliferation, 7, 1, 2, 5, 9, 11, 14, 21, 22, 24, 31, 36
promote, 32
property, iv
protection, 14
prototype, 24
public, 16, 42

R

radiation, 7, 14, 23
radioactive waste, 7, 1, 2, 6, 8, 10, 12, 15, 16, 22, 27, 42, 44
range, 19, 20
reasoning, 37
recycling, 7, 1, 2, 3, 6, 7, 8, 9, 10, 11, 12, 13, 14, 15, 17, 19, 21, 22, 23, 24, 25, 26, 27, 28, 31, 33, 35, 39, 40, 41, 42, 43, 44, 45
reduction, 1, 2, 11, 12, 13, 20, 21, 22, 27, 29, 31, 35, 39, 41
regulations, 10
reliability, 16
reprocessing, 1, 3, 6, 9, 10, 11, 13, 14, 15, 16, 17, 18, 19, 22, 23, 24, 26, 27, 28, 32, 33, 35, 36, 40, 41, 44, 45
research, 10, 35, 39
Research and Development, 1, 5, 8, 9, 11, 13, 14, 16, 17, 18, 19, 22, 23, 24, 27, 31, 32, 33, 36, 39, 40, 41, 42, 44
researchers, 40
reserves, 5
resistance, 9, 36
resources, 27, 31, 33, 35, 44
risk, 1, 5, 7, 14, 17, 18, 24, 28, 31, 32, 43
risks, 7, 1, 2, 11, 21, 31, 36, 43

S

safeguards, 14, 16, 19, 23, 32, 33, 40, 45
safety, 10, 19, 20
sales, 25, 26
Savannah River Site, 14, 16, 17, 18, 40
scaling, 17
scheduling, 16
security, 5, 14
Senate, 37, 38
separation, 15, 16
services, iv, 28
sharing, 25
sodium, 10
speed, 2, 24
spent nuclear fuel, 7, 1, 2, 5, 6, 7, 10, 12, 13, 14, 15, 26, 28, 31, 33, 36, 39, 40, 41, 42
stages, 28
standards, 14, 42
storage, 8, 22, 36
strategic, 7, 2, 22, 24, 25, 28, 31, 36, 41
strontium, 8, 15, 16
surplus, 26
systems, 40

T

technology, 7, 5, 9, 10, 14, 16, 17, 22, 24, 26, 27, 28, 37, 40, 43, 44, 45
temperature, 6, 12, 15, 22
theft, 9, 14, 24, 28
time, 6, 10, 17, 23, 25, 27, 33, 37, 44
time frame, 17, 25, 27, 33

timing, 31
transition, 2, 29, 31, 36

U

United States, 9, 10, 19, 23, 26, 28, 37, 38, 45
uranium, 2, 5, 7, 8, 9, 17, 22, 23, 28, 37
uranium enrichment, 28
uranium oxide, 9

W

waste management, 1, 15
wastes, 10
water, 10, 12, 13, 22, 24, 25, 26, 28, 35, 44
weapons, 2, 36
workers, 19

Y

Yucca Mountain, 6, 15, 16, 42